Genre Nonfiction

Essential Question

What are maglev trains?

Maglev Trains

by Andrew Seear

Chapter 1
Are We There Yet?

How many times have you asked the question "Are we there yet?" You probably ask it because you want to get somewhere quickly.

Today, people travel long distances for business and fun. They want travel to be safe and comfortable. No one wants to sit in traffic or wait for a flight that is late. These things can add hours to a trip. The price of fuel also affects the cost of all types of travel.

Is there a way to travel that saves gasoline, is safe, is comfortable, and does not cost a lot of money? Some people say there is— the maglev train.

The *mag* in *maglev* is short for **magnetic**. The *lev* is short for **levitation**. A maglev train is a train that floats in the air above a track. It is held up by the force of **magnets**. Magnets not only make the train float, but they also make it move. Electricity gives the magnets energy to do their work.

This maglev train floats above the track.

Best View Stock/Alamy

An Untestable Idea

About a hundred years ago, Robert Goddard and Emile Bachelet both had ideas for magnet-powered transportation. But at the time, there were no magnets powerful enough to test their ideas.

In the 1930s, Hermann Kemper in Germany also had an idea for magnet-powered transportation. Yet there were still no magnets powerful enough to test his design.

Then, in 1968, two Americans, James R. Powell and Gordon T. Danby, designed a magnet-powered train. By this time, super-powerful magnets were in use. The maglev train finally became more than an idea.

Robert Goddard

NASA

"Full Steam Ahead"

Soon, many countries were interested in building a maglev train. Scientists designed and tested many models. This was very costly. The United States government decided working on a maglev train cost too much money. The project ended in 1973.

Other countries kept working. In 1991, Germany's maglev train was ready for use. It was called *Transrapid*. It was very successful. As a result, the United States started working on a maglev train again.

An early maglev train was tested in 1973.

Chapter 2
How a Maglev Train Works

In some ways, a maglev train looks like a regular train. There is a cabin at the front where the driver sits. Behind it, there are train cars for the passengers. The maglev train has rails to guide it. It stops at stations so that passengers can get on and off.

A maglev train at a station

There are also many differences between a maglev train and an ordinary train. A maglev train has a driver, but no engine. It has steel **guideways** that look like rails, but it does not run on the rails. It floats above them.

The maglev train uses magnetic forces to levitate. The steel guideways of the trains have coils inside. They are called levitation coils, or levitation magnets. These coils are found all along the length of the guideway.

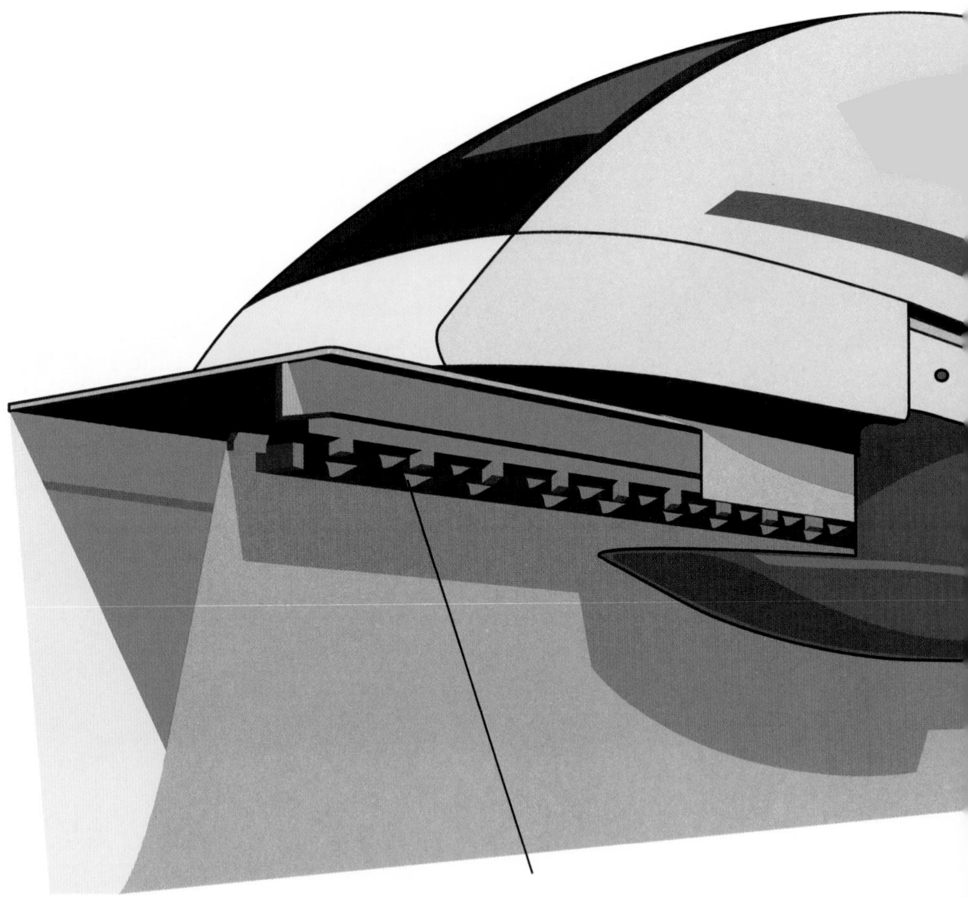

These guideway magnets push and pull the train, moving it backwards and forwards.

On the bottom of the train itself, there are also very powerful magnets. When the train is moving, these magnets pass by the coils in the guideway. Electricity is sent through the coils. They are **electromagnets**. The magnetized coils push or pull the train up off the rails. The same forces move the train forward. They push and pull the train along.

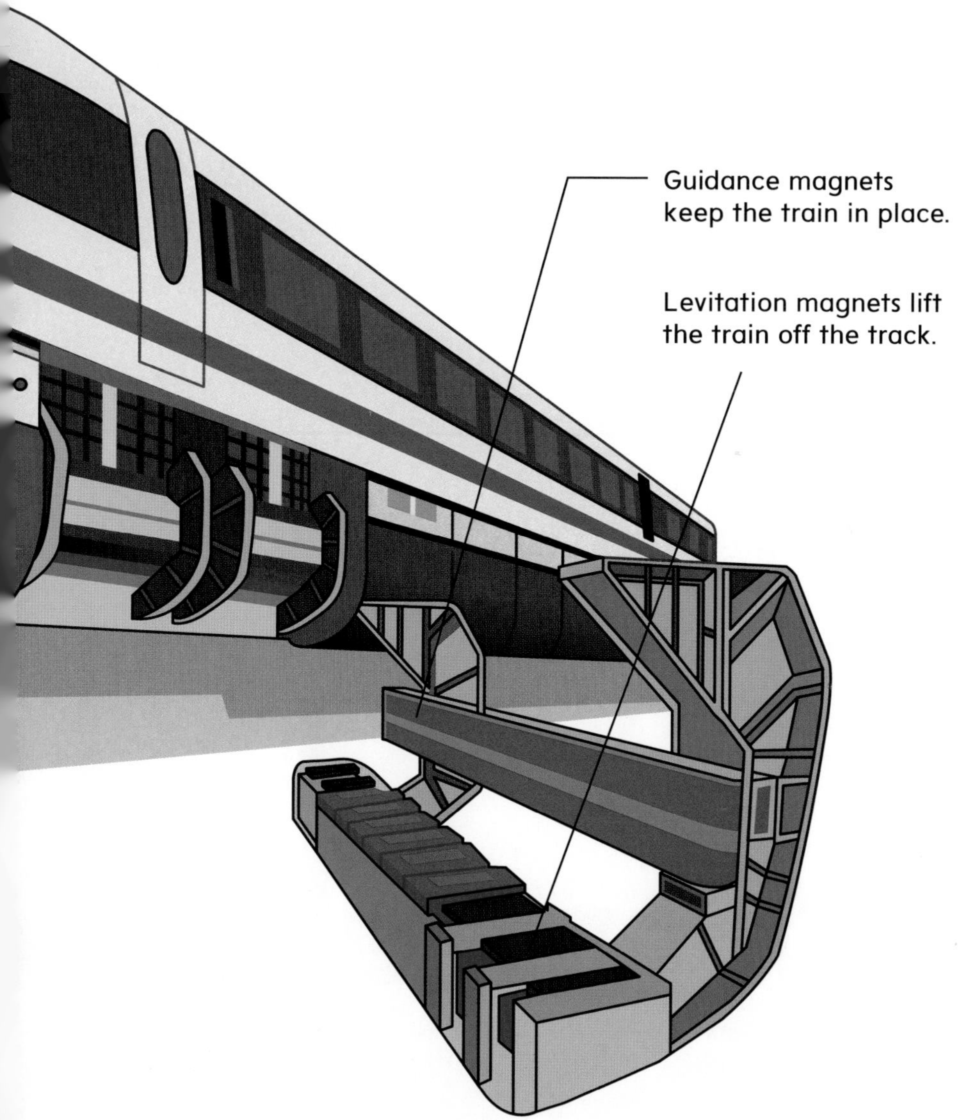

Guidance magnets keep the train in place.

Levitation magnets lift the train off the track.

Chapter 3
The Good and Not So Good

Maglev trains are the fastest trains in the world. The fastest train in the United States is the Acela. It can travel as fast as 241 kilometers (150 miles) an hour. Maglev trains can travel at speeds twice as fast. They can travel more than 482 kilometers (300 miles) an hour!

Maglev trains do not need repairs as often as airplanes, cars, and other trains. They are much easier to keep running. They are also safer than other kinds of transportation.

Electric current in overhead wires helps make this railroad train move.

Maglev trains do not use gasoline. They do not **pollute** the air, and they make much less noise than cars, airplanes, and other trains. A maglev train does not harm the **environment**.

Maglev trains can move large numbers of people from place to place. They can also move freight, or goods, that are usually carried by trucks or railroad trains. If people and freight traveled by maglev train, that would mean fewer cars and trucks on the road.

Maglev trains are quieter than railroad trains.

A Smooth Ride

Maglev trains do not run on steel rails, so the ride feels smooth. One type of maglev does not have wheels, so no wheels hit the track. There is no **friction** to slow down the train or cause **vibration**. They do not rattle like ordinary trains on a track. A ride on a maglev is smooth and comfortable.

Passengers can feel the difference.

Fat/Alamy

Concerns about Maglev Trains

Maglev trains and guideways are expensive to build. Experts figured out the cost of building a maglev line from downtown Washington, D.C., to Baltimore, Maryland. The 59.5-kilometer (37-mile) trip might cost between 8 billion and 10 billion dollars. That's nearly 300 million dollars per mile!

This map shows where a maglev train route might be built.

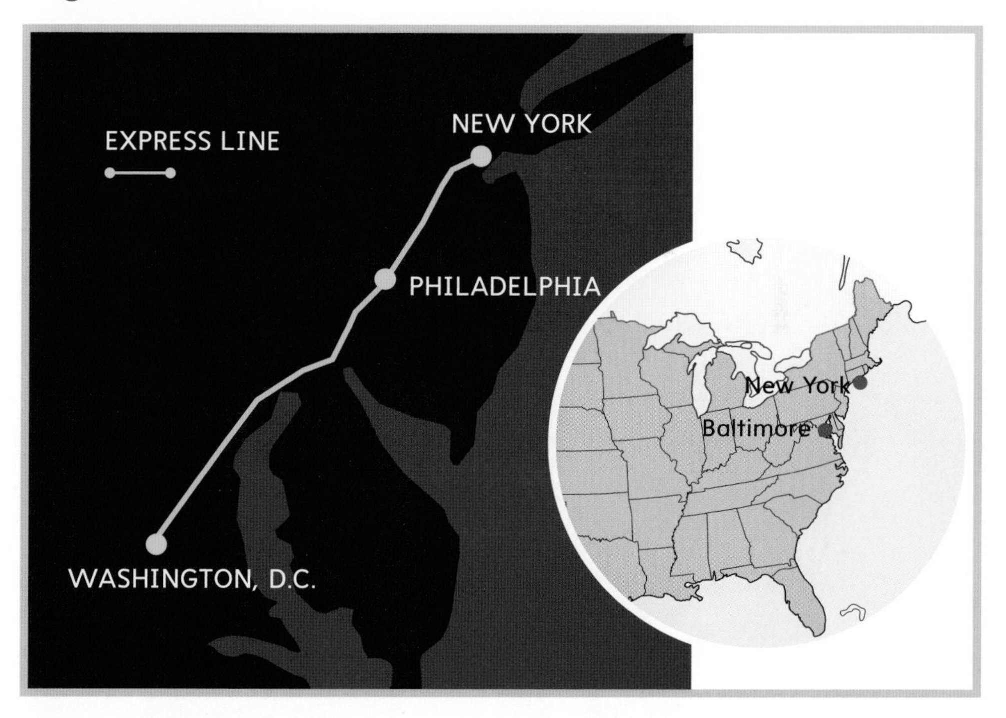

It will also not be easy to connect maglev trains with ordinary railroad trains. Maglev trains cannot run on railroad tracks. New stations will have to be built for those trains.

Building guideways will also mean changing the environment. Trees will have to be cut down. Soil will need to be dug up. Rocks will need to be moved out of the way, to make room for maglev guideways.

Many trees have to be cut down to make space for maglev guideways.

Maglev trains run best on long nonstop rides between cities. They are not like ordinary trains. Ordinary railroad trains can stop and start more easily. They can pick up and drop off passengers at smaller stations between cities.

Railroad trains can stop easily and in more places.

Christian Petersen-Clausen/Moment Open/Getty Images

Maglev trains do not start and stop easily. It takes time for a maglev train to reach its maximum speed. One of the main benefits of a maglev train is its pace. If the maglev train keeps starting and stopping, it will not reach its high speeds.

Maglev trains travel too fast to make many stops.

Chapter 4
The Future of Maglev Trains

In spite of the problems and costs, many countries are going ahead with maglev projects. Japan is building a maglev line between Tokyo and Nagoya, two major cities.

This maglev route in Japan will allow people to travel between two cities quickly.

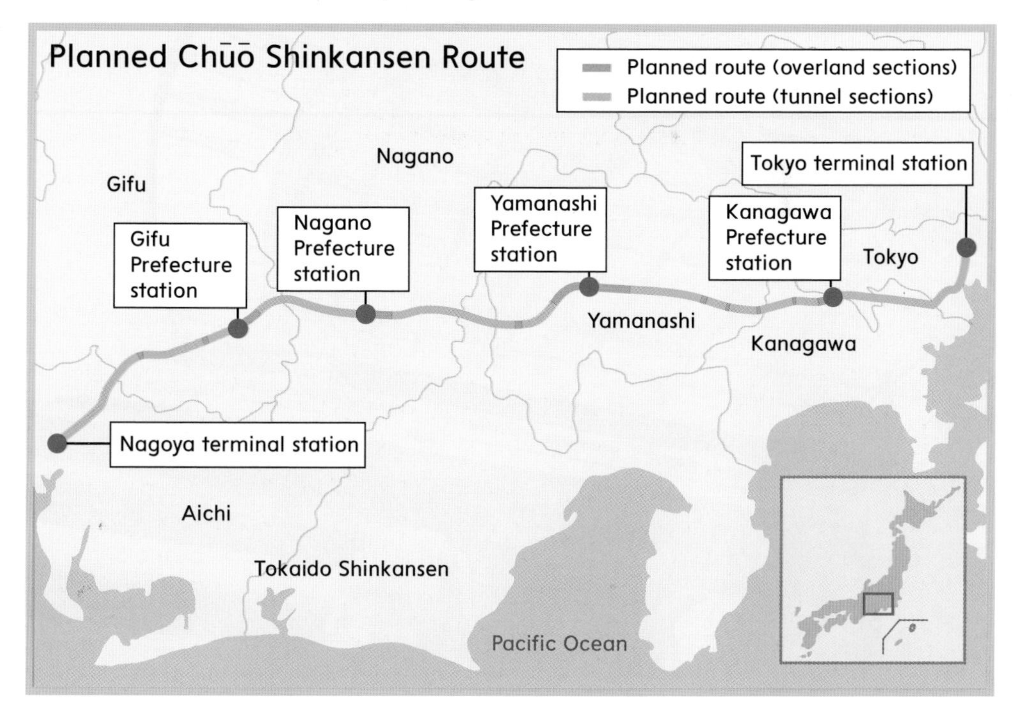

A maglev train began to operate in China in 2004. It covers the 30 kilometer (19 mile) trip between Shanghai airport and the city center in less than ten minutes. The ride is short. Yet it proves that maglev trains can be a fast, safe, comfortable, and easy means of transportation.

This train speeds out of the station.

Nikada/Getty Images

The United States is working on maglev projects, too. The California-Nevada Super Speed Train Project is one example. Another route is planned between Washington, D.C, and New York City. There is a project in the planning stage in Pennsylvania. A maglev route between Chattanooga, Tennessee, and Atlanta, Georgia is also planned.

Be on the lookout for maglev trains. One may be coming your way soon!

New maglev train routes are being planned in the United States.

Respond to Reading

Summarize

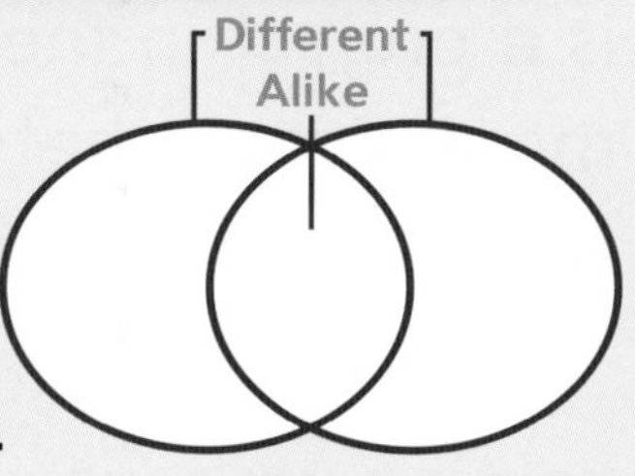

Use important details from *Maglev Trains* to summarize the selection. Your graphic organizer may help you.

Text Evidence

1. How can maglev trains help solve some of the world's traffic problems?
2. Reread the book with a partner. Work together to make a Venn diagram. In the outer circles, list the differences between a maglev train and an ordinary train. In the center circle, list how the two types of trains are alike. COMPARE AND CONTRAST
3. What is the meaning of the word *floats* on page 3? What else could the word *floats* mean? What clues in the text show you which meaning to use on page 3? HOMOGRAPHS
4. Write a letter to your mayor. Ask him or her to consider building a maglev train in your city. Be sure to explain the benefits a maglev train would bring. Compare them to an ordinary train. Use strong reasons to persuade. WRITE ABOUT READING

Genre Fiction

Compare Texts

Read about how girl and her family experience science and technology as they ride on a very fast train. All aboard!

Smooth as Air

"Mom! When will we finally arrive in Shanghai?" Annchi dad chuckles at her and politely requests quietly, "Maximum patience please!" They tiredly trudged into a shiny train station. Annchi despairingly grumbled, "No, more waiting! Train travel takes forever! We'll never get to Shanghai!"

The long white train came in with a deep rumble. It resembled the end of a thunderclap! "Dad!" she screamed, "The wheels are missing! It's levitating above the track! Everyone aboard without delay!"

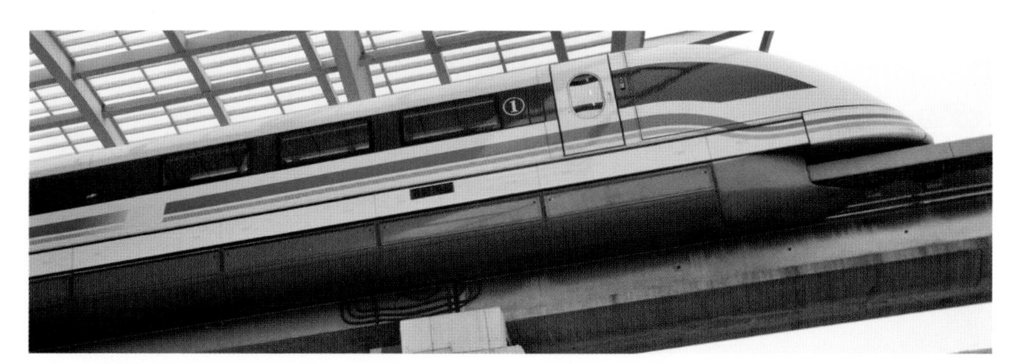

Nikada/Getty Images

Annchi is totally amazed at what she finds inside the train. The seats are very comfortable after a long day. She soon observed a digital time and speed monitor. She glued her eyes to the monitor's illuminated, ever changing data. Time 30 Seconds Speed 60 mph. Time 45 s. Speed 120 mph. Wow! Objects outside the train are blurry. "Outstanding!" Time 1 min. 15 s. Speed 200mph! Time 2 min. Speed 300 mph "This is incredible!"

Annchi yelled. A train shrieked rapidly past them in the other direction!

The train stops very slowly. Her dad says, "Twenty miles in eight minutes. Not bad!" She laughed, "I want to experience that again! Right now!"

Sergiy Serdyuk/Hemera/Getty Images

Make Connections

What is the science that lets maglev trains go much faster than other trains?

TEXT TO TEXT

Glossary

electromagnet *(i-LEK-troh-MAG-nit)* a temporary magnet formed when current flows through wire wrapped in coils around an iron bar *(page 9)*

environment *(en-VIGH-ruhn-muhnt)* the things that make up an area, such as land, water, and air *(page 11)*

friction *(FRIK-shuhn)* a force between surfaces that slows objects down or stops them from moving *(page 12)*

guideway *(GIGHD-way)* a steel track that has magnets and controls the movement of a maglev train *(page 7)*

levitation *(lev-i-TAY-shuhn)* the act of rising and floating in the air *(page 3)*

magnet *(MAG-nit)* any piece of iron or certain other metals that has the property of attracting iron or some other metals *(page 3)*

magnetic *(mag-NET-ik)* having the properties of a magnet *(page 3)*

pollute *(puh-LEWT)* to add harmful things to the water, air, or land *(page 11)*

vibration *(VIGH-bray-shuhn)* the act of moving back and forth *(page 12)*

Index